AUSTRALIA'S MOST
FREAKY

—

By Karen McGhee

Australian
GEOGRAPHIC

AUSTRALIA'S MOST
FREAKY

First published in 2014 by:

BAUER
MEDIA GROUP

Bauer Media
54 Park Street, Sydney, NSW 2000
Telephone (02) 9263 9813
Email editorial@ausgeo.com.au

www.australiangeographic.com.au

Australian Geographic customer service:
1300 555 176 (local call rate within Australia).
From overseas +61 2 8667 5295

Text Karen McGhee
Editor Averil Moffat
Book design Mike Ellott, Mike Rossi
Picture research Jess Teideman
Print production Chris Clear
Education Coordinator Lauren Smith
Sub-editor Amy Russell
Managing Director Matthew Stanton
Associate Publisher, Specialist Division Jo Runciman
Editor-in-Chief, Australian Geographic Chrissie Goldrick

National Library of Australia
Cataloguing-in-Publication entry:

McGhee, Karen, author.
Australia's most freaky: weird and wonderful creatures / Karen McGhee.

ISBN 9781742455068 (paperback)

Animals – Australia – Juvenile literature.

590.94

Printed in China by C & C Offset Printing Co. Ltd.

OTHER TITLES IN THIS SERIES:

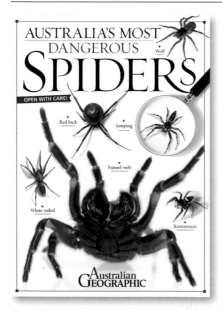

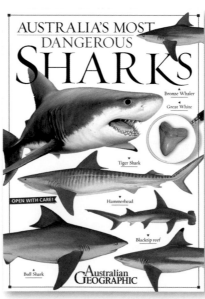

AUSTRALIA'S MOST FREAKY

WEIRD AND WONDERFUL CREATURES

Learn about some of Australia's most intriguing and wonderfully weird animals. Explore these pages to discover the bizarre body shapes and colours, extraordinary lifestyles and amazing hunting techniques of our freaky creatures.

CONTENTS

Q&A

Q: Will platypus venom kill you?
A: No. But it's said to be one of the most excruciatingly painful venoms.

MAMMALS THAT LAY EGGS

These mammals are Australian icons, but they're also among the weirdest creatures on the planet.

Platypus

It has fur and produces milk to feed its young, so it's a mammal. But it lays eggs like a reptile. That strange combination is freaky enough. The platypus also has a bill shaped like a duck's, a tail like a beaver and webbed feet. But the platypus also produces venom. This comes from spurs on the males' hind legs that are connected to venom glands.

FACT BOX

Semi-aquatic lifestyle

Platypuses are very well adapted to life in and around rivers and streams. Their webbed front feet and partly webbed back feet help them swim underwater as well as scamper across river banks. They have sturdy claws on their front feet for digging burrows where they shelter and raise young. And their fur is extremely thick and waterproof.

Did you know?

When scientists in Europe in the late 1700s first saw a platypus skin they thought it was a forgery and looked for stitches connecting its rubbery bill to its fur.

Short-beaked Echidna

Apart from the platypus, echidnas are the only other egg-laying mammals. The short-beaked echidna is found right across the Australian mainland and on many surrounding islands. Echidnas are adapted perfectly to a diet of ants and termites. Their short powerful front paws have long, sharp backward-curving nails that they use to break into termite mounds and ant nests. Their long, thin tongue is sticky, which helps them catch lots of ants at once.

What do you call a baby echidna?

Echidna babies are known as puggles. Like platypus babies, they are born very underdeveloped; blind, hairless, defenceless and totally dependent on their mothers for the first months of life.

FREAKY FACT

MY SPIKES ARE MADE OF HAIR!

FACT BOX

Echidna defence

Echidnas have a layer of short coarse fur that insulates their bodies against the weather and a longer outer layer that forms tough spines. When threatened, echidnas protect their soft underbellies by digging into the ground or curling into a ball leaving only their spiky spines exposed.

SPECIALISED MAMMALS

The most famous Australian animal group is the marsupials – mammals with pouches for developing babies. But built-in maternity sacs are just one of many weird features found in our mammal species.

Greater bilby

It's thought that the huge, near-naked ears on this desert marsupial help it release body heat. The large size and shape of their ears also ensure that bilbies have excellent hearing. They need to because, to avoid high daytime temperatures, they **forage** only at night. Along with a great sense of smell, exceptional hearing helps bilbies locate the bugs and grubs that form an important part of their diet.

Spinifex hopping mouse

This little rodent has a range of adaptations that help it survive the intense heat and arid conditions of Australia's deserts. For one, it produces the most concentrated urine of any mammal in the world. In fact, its pee is so concentrated that it's almost solid. The spinifex hopping mouse has highly specialised kidneys and can survive very long periods without the need to drink water. It gets all the moisture it needs from its diet.

Did you know?

Bilbies are spiral diggers. They use their strong front limbs to dig deep twisting burrows that extend underground for more than 2m. The corkscrew shape helps stop predators, such as goannas, from digging them up.

Marsupial mole ▶

Australia's marsupial moles are among the best mammal burrowers in the world. They live in deserts and rarely appear above ground. Rather than dig tunnels, like most other mammals that live underground, they 'swim' through sand. They lead with their calloused nose and forehead and follow through with spade-shaped front feet powered by well-developed shoulders.

FREAKY FACT

MY EYES ARE COMPLETELY USELESS!

Marsupial moles are blind, their ears are very small and they have fine silky fur that's well-suited to gliding through their closed-in sandy habitat.

Deadly dilemma

All killer diseases are nasty. But there's one that's threatening to wipe out Australia's Tassie devils – a kind of cancer known as Devil Facial Tumour Disease, or DFTD. It spreads between the animals when they fight. In some devil populations, it's killing more than 90% of adults and the Tasmanian devil has now become endangered.

Tasmanian devil

Early European settlers gave the devil its common name because of the eerie growls and screeches it makes at night when it's fighting over food and territory. It once lived right across Australia but is thought to have become extinct on the mainland at least 400 years ago, before Europeans arrived. This was probably caused partly because of competition from dingoes. Devils are now only found naturally in Tasmania.

FLYING MAMMALS

From carnivores with wings to fruit-eaters with noses like hoses –
life in the air doesn't get much battier.

Eastern tube-nosed bat

Large tubular nostrils that bulge from the sides of the face are the most distinctive feature of this bizarre-looking rainforest creature. The main component in its diet is fruit. And it was once thought their weird noses functioned like snorkels, allowing them to breathe as they buried their faces in soft ripe fruit while eating. But scientists have discovered that this is not the case. ▼

Ghost bat

These tropical night-time fliers are Australia's only meat-eating bats. They drop down from the sky onto prey, wrap their thin leathery wings about it and kill by biting its head and neck. Animals that ghost bats attack and eat in this way include small mammals, birds, frogs reptiles, large insects and even other bats. Despite their gruesome lifestyle, they're secretive and very nervous creatures that are easily disturbed from the caves where they roost by day. When not hunting they often hang about in small colonies.

Flying spooks
The fur of these bats is coloured in shades of grey and some are so light that they look almost ghostly white as they flit about in the moonlight.

Knowing noses
The nostrils of eastern tube-nosed bats can open, close and move independently of each other, which must give these bats a great ability to locate ripe fruits.

Q&A

Q: What do you call an animal that specialises in eating fruit?
A: A frugivore.

MARINE MISFITS

Big doesn't necessarily mean beautiful!

Dugong

These massive creatures are like the ocean's version of cows. They are plant-eaters that graze on the meadows of seagrass found in shallow coastal waters. Most of the world's dugongs live in the tropical waters of northern Australia. Their distinctive facial features make them easy to identify. They have elongated, flattened and down-turned snouts with a huge bristly upper lip, perfectly suited to 'mowing down' underwater plants. It's said that dugongs may have inspired tales of mermaids, with sailors mistaking them for women – but that's very hard to understand!

Did you know?

Dugongs are more closely related to elephants than they are to marine mammals such as whales and seals.

Southern elephant seal

Male southern elephant seals are truly enormous sea creatures: males can reach a length of 4m and weigh 3.5 tonnes. But size isn't the only reason for their common name. The males have a huge, trunk-like nose. It's used to make loud roars on land to impress females and warn off other males during the breeding season.

FREAKY FACT
THE GIRLS FIND MY NOSE ATTRACTIVE!

Champion divers

Elephant seals can hold their breath for up to two hours and have been recorded down to a depth of 2km. No other seal in the world can match that.

SOUTHERN CASSOWARY

This odd-looking bird is a shy forest creature with a kick that can pack a deadly punch.

Cassowary

It's got a long neck and head covered in mostly blue, lizard-like scaly skin. And a pair of long, red skin flaps called 'wattles' hang from its throat. But one of the Australian cassowary's weirdest features is the large grey helmet-like lump, called a 'casque', that pokes out of its head by as much as 18cm. It's made of toughened skin, hard on the outside but spongy inside.

FACT BOX

Devoted dads

Females have nothing to do with their eggs once they've laid them. The males **incubate** and hatch them and then look after the chicks for more than a year. The chicks' fluffy stripes are a type of camouflage known as 'disruptive colouration'. It makes their shape seem to disappear in dense undergrowth.

BIRDS WITH SPECIAL ABILITIES

Unique talents set these birds apart.

Letter-winged kite

This is the world's only truly **nocturnal raptor**. Nocturnal hunting makes good sense in the arid and desert areas where it lives as most animals avoid high daytime temperatures and only come out at night.

Palm cockatoo

Male palm cockatoos break twigs off trees and use them like drumsticks to beat on nesting trees, perhaps to mark territory. Making and using tools is rare in the animal kingdom. But using a tool for something other than collecting food is almost unheard of.

Malleefowl

Instead of building nests from sticks and twigs, malleefowl scrape together huge mounds of dirt and incubate their eggs inside. The male maintains the mound temperature at a constant and perfect 33°C, mostly by scraping away or applying more dirt or rotting vegetation as needed. He tests the temperature by sticking his head inside and poking his tongue around. There are special heat sensors on his tongue.

FREAKY FACT

MY NEST CAN WEIGH UP TO 300 TONNES!

OUTRAGEOUS SHOW-OFFS

Some Australian birds have the weirdest ways of strutting their stuff.

Superb lyrebird

Australia's superb lyrebird is the master mimic of the bird world. It makes some of its own clicks and whistles, but most of its 'songs' are based on other sounds it hears. That includes the songs of other birds in eastern Australia's bushland, from the raucous laughter of kookaburras to the ringing melodies of bellbirds.

FREAKY FACT

I CAN COPY LOTS OF SOUNDS!

FACT BOX

Male show-offs

An adult male lyrebird is just a drab brown bird until he raises his tail. When his lacy tail feathers are spread, the tail looks like the ancient Greek musical instrument called the lyre, which is how these birds got their name. The stunning feathers can measure almost one metre, longer than the rest of their body.

Lyrebirds don't just mimic other birds. They also copy other animals as well as sounds made by machines, like the revs of chainsaws or dirt bikes. Their songs can also include perfect renditions of honking car horns, screeching alarms, whirring camera shutters, the cries of babies and the barking of dogs.

Q&A

Q: On what Australian coin does the superb lyrebird appear?
A: The 10-cent coin.

Satin bowerbird

These birds get their common name because the males build ground structures called bowers. These are usually parallel rows of sticks separated by a cleared space. All male bowerbirds use their bowers as a stage for complex courtship behaviour. This includes a strange obsession. They collect and display bright blue objects as a way to woo prospective mates.

FREAKY FACT

MY FAVOURITE COLOUR IS BLUE!

In the bush, male satin bowerbirds use mostly blue feathers and flowers to decorate their bowers. But if human settlements are near, these birds will plunder backyards for anything blue – from clothes pegs to drinking straws. Blue bottle tops seem to be highly prized.

FACT BOX   Performance spaces

Males build bowers to impress females, they're not for nesting. A female often visits on her own first and, if impressed, she'll return for the male's courtship dance. It's an energetic performance of strutting, quivering and bowing, usually with one of his blue objects held in his mouth. If the show works, the pair mate in the bower.

TURTLES WITH TALENT

The name says it all as far as these turtles are concerned.

FREAKY FACT

I CAN BREATHE THROUGH MY BUM!

The scientific term for 'bum breathing' is 'cloacal respiration'. A small number of other turtles are known to do it but the Fitzroy River turtle does it best.

Fitzroy River turtle

This is the polite common name for this turtle. It's more widely known as the 'bottom breathing turtle' or even the 'bum breathing turtle'. And these names describe exactly what it's famous for; breathing through its back passage. In turtles, as for other reptiles, that passage is called a 'cloaca'. Both urine and faeces, as well as eggs pass out through here.

Pig-nosed turtle

It's obvious how this turtle got its name. Its nose is large, fleshy, pig-like and it's perfectly adapted to the lifestyle of this turtle, which spends almost all of its time underwater. Its odd-looking nose works just like a snorkel: it lets the turtle breathe at the water's surface while the rest of its body remains submerged.

Thorny devils are only about 20cm long and are harmless to humans. But they're scary looking lizards. They're covered in thorny scales that help deter predators. The scales also help them extract fluids from their desert habitat. Narrow channels running between the scales towards the mouth act like straws to 'suck' water from dew-covered plants.

LOOKS CAN BE DECEIVING

These lizards seem a lot scarier than they really are.

Frill-necked lizard

This lizard is all bluff and no bite. When threatened it throws its head, opens its bright yellow-coloured mouth widely and lets out a hiss. For even greater effect it flares out a spectacular frilly fold of skin from around its neck. It can measure 30cm across when fully opened. This is all meant to make the frill-necked lizard look larger and fiercer than it really is.

Blue-tongue lizard

Small legs; round, solid bodies; and a large triangular head: blue-tongue lizards are definitely not built for speed. And so, when threatened, they stand their ground and try to bluff their way out of it. They hiss and flick out their tongue, which is frighteningly blue against the bright pink of the rest of the mouth.

Death adder

Most experts put these snakes on the world's top 10 list of deadliest reptiles. They also have the longest fangs of any Australian snake. They are **ambush** predators, meaning they don't hunt for food. Instead they hide and wait for small mammals, lizards or even other snakes to cross their path. Then they strike with lightning speed. Also, adders have a growth on the end of their tail that looks like a worm. They twitch this about like a lure, to tempt prey within reach.

EXTREME SERPENTS

Australia has more deadly snakes than any other country! But deadly venom isn't all there is to these freaky serpents.

FREAKY FACT

AUSTRALIA'S LONGEST SNAKE!

Inland taipan

There's no hiding it! Yes, Australia is home to the land snake with the deadliest venom and this is it. A bite from the inland taipan can kill an adult in less than an hour. The good news is it's rare and lives in remote habitats in arid Australia.

Amethystine python

This python is Australia's longest snake, although it's not quite the world's longest. That honour goes to South-East Asia's reticulated python, which can reach about 10m. The amethystine python's maximum length is about 8.5m, which is still a pretty freaky length for a snake!

PREHISTORIC CREATURES

Australia's crocodiles haven't changed much since the dinosaurs, more than 65 million years ago.

Saltwater crocodile

No other living reptiles get as big as Australia's saltwater crocodiles. Adult males can grow longer than 7m and weigh more than a tonne; as big as the average family car!

FREAKY FACT

THE BIGGEST REPTILE IN THE WORLD!

Although crocodiles breathe air, special adaptations allow them to remain underwater for more than an hour without coming to the surface. This means they can lurk silently just below the surface of rivers and creeks, ready to explode out and grab prey that comes too close.

Freshwater crocodile

This croc doesn't get nearly as big as its saltwater relative – usually no longer than 2m. And they are not nearly as powerful or aggressive as 'salties'. Most importantly, their jaws and snout are a lot smaller and thinner.

FABULOUS FROGS

Many of Australia's frogs are the weirdest amphibians you'll find anywhere on the planet.

FREAKY FACT

I CAN SURVIVE YEARS WITHOUT WATER!

Desert trilling frog

In Australia's deserts there can be years when there's no rain, then suddenly there's a flood, then it's back again to years of drought. These are called 'boom–bust cycles' and they suit desert trilling frogs just fine. These frogs are so well adapted to desert life that they can survive deep underground in burrows for years. They store water in glands beneath their skin and protect themselves inside cocoons made of old skin cells until rain comes.

Corroboree frog

These striking frogs are found only in Mount Kosciuszko's alpine marshlands. They **hibernate** through the winter. The bright yellow-green and black markings tell potential predators these frogs are poisonous and best left alone. Many frogs have poisons in their skin and they get these through the food they eat, such as ants. But corroboree frogs are the first to be discovered that make their own.

Gastric-brooding frog

No other frogs take care of their babies in the way these unique Queensland frogs do: by swallowing them! After her eggs have been fertilised, the female gastric-brooding frog gobbles them down. They hatch in her stomach and develop into tadpoles. About seven weeks later she spits out fully developed young frogs.

FREAKY FACT

I SWALLOW MY BABIES! GULP!

Marsupial frog

Males of this frog species have pouches where they carry tadpoles. After mating, the female marsupial frog lays her eggs onto leaf litter and moist soil. She stays and watches over them for a few days until they are ready to hatch. Then the male moves in, lies over the eggs and the newly hatched tadpoles wriggle up into pouches on his hips. He carries about six tadpoles on each side and they stay there until they develop into small frogs.

Turtle frog

This bizarre Western Australian frog looks just like a turtle without its shell. It's just 5cm across, round and pink. It has tiny eyes, a small head compared to its body and strong front limbs that it uses to burrow headfirst into the sand. That's the opposite way to most burrowing frogs, which dig backwards into the soil using their hind legs.

Crucifix frog

Frogs breathe through their skin so they need to keep it in perfect condition. It produces a range of different chemicals for different reasons. Some keep viruses and bacteria away. Some are poisons that are meant to put predators off. But crucifix frog skin produces something weird, wonderful and unique; a sticky, elastic 'glue'. No one is exactly sure why crucifix frogs produce this 'glue', but scientists are looking to see if they can develop it for use in human medicine.

FREAKY FISH
STRANGE

Confounding camouflage helps to hide these Australian fish.

Reef stonefish

These fish live on tropical and sub-tropical reefs along Australia's coastlines. There they blend in perfectly by looking just like lumps of coral or rocks covered with algae. By keeping very still, they **ambush** passing prey.

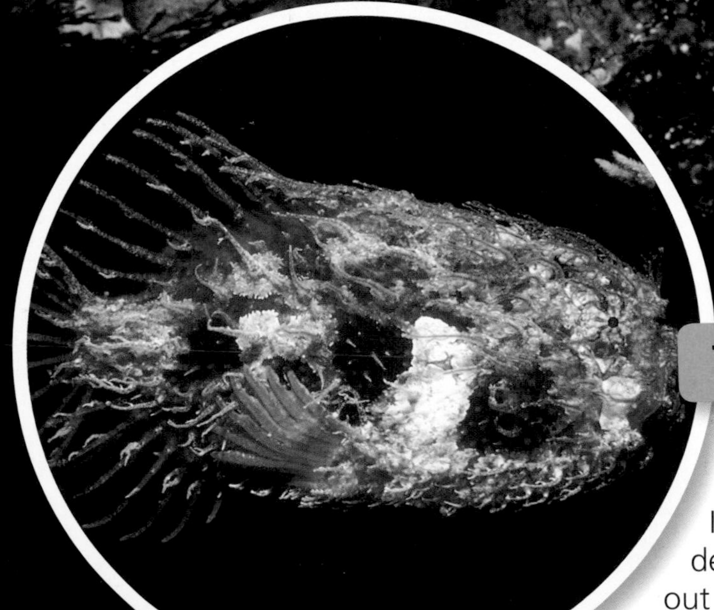

Tasseled frogfish

The tasseled frogfish is motionless as it waits for prey in cluttered sponge gardens. It blends in superbly, hidden by its own dense covering of fleshy filaments that grow out from its body. This clever fish doesn't rely only on chance meetings with passing prey. It carries its own 'fishing rod' that extends out from its dorsal fin. Hanging on the end, and positioned not far from its mouth, are two growths that it can wiggle just like worms.

FREAKY FACT

I PACK MY OWN FISHING ROD!

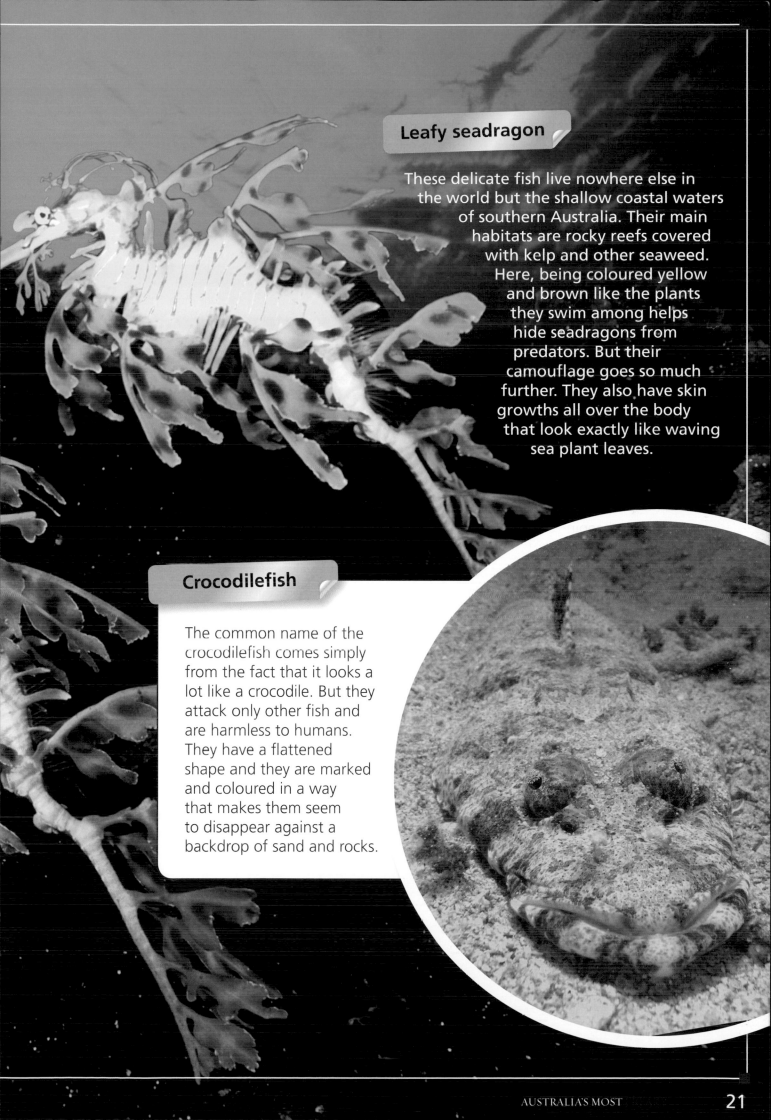

Leafy seadragon

These delicate fish live nowhere else in the world but the shallow coastal waters of southern Australia. Their main habitats are rocky reefs covered with kelp and other seaweed. Here, being coloured yellow and brown like the plants they swim among helps hide seadragons from predators. But their camouflage goes so much further. They also have skin growths all over the body that look exactly like waving sea plant leaves.

Crocodilefish

The common name of the crocodilefish comes simply from the fact that it looks a lot like a crocodile. But they attack only other fish and are harmless to humans. They have a flattened shape and they are marked and coloured in a way that makes them seem to disappear against a backdrop of sand and rocks.

DIFFICULT HABITATS

The aquatic world contains many freaky creatures too. Weird shapes, strange habits, amazing lifestyles – they're all found underwater.

Kerryn Parkinson/© NORFANZ Founding Parties

Blobfish

Ever since the early 2000s when it was first brought up in trawls from the deep ocean off south east Australia, this blobfish has been famous for being ugly. But there's good reason for that unusual blobby appearance. It's that way because its habitat is literally very high-pressured.

FACT BOX

A perfect body

Down at 800m where blobfish live, the surrounding pressure is 80 times what it is at the surface. The blobfish doesn't need bones to support it: in fact, they'd be crushed. And it doesn't need muscles because it has no need to move far or fast. Instead, it's made of mostly jelly-like tissue that's slightly less dense than the surrounding seawater. And that helps it to float perfectly above the sea floor, where it sucks up passing food particles. In their own deep-water high-pressure environment blobfish look more like normal fish.

Australian lungfish

Most fish can only breathe underwater. Their gills extract oxygen from water. Out of the water they quickly die. The Australian lungfish is one remarkable exception. It's able to breathe oxygen from air and can survive days out of water if it's kept moist.

Elephantfish

It's obvious how this fish gets its common name. The club-like structure on its snout looks like an elephant's trunk. This weird-looking body part is used to scan the ocean's floor for prey. Its end is covered in pores that can detect the faint pulses of electricity given off by many marine animals. Elephantfish are chimaeras; a type of fish closely related to sharks and rays and, like them, they don't have hard bones. Their skeletons are made of cartilage – a firm but flexible material.

Did you know?

Elephant fish are caught commercially in the waters off southern Australia and New Zealand, where they live. Their fillets often end up being sold for use in 'fish and chips'.

ODD APPENDAGES

Strange-shaped body parts have a purpose.

Spotted handfish

These highly unusual fish don't swim. Instead they get around on the ocean floor by 'walking'. And they do this by using their pectoral fins, on the sides of their bodies, which have become modified through evolution and now look like 'hands'. The spotted handfish has so far only been found in the Derwent River, near Hobart, Tasmania.

GATHERINGS

A lot of animals group together to breed, but these aggregations are over the top.

Christmas Island red crab

This land crab is found only on Christmas Island, in the Indian Ocean off Western Australia. About 120 million of them live on the island, most in burrows on the floors of rainforest habitat. Towards the end of each year, at the beginning of the wet season, most adult crabs head towards the sea on a migration that takes more than two weeks.

Australian giant cuttlefish

Each year in late autumn, hundreds of thousands of these huge cuttlefish gather together to mate in a small ocean inlet off South Australia called Spencer Gulf. These creatures only live for a year or two and die soon after their first and only mating. Cuttlefish are 'cephalopod molluscs', a group that includes squid and octopus. All of these creatures are able to change their body colour, pattern and shape. The giant cuttlefish is an expert and can dramatically change its appearance in less than a second.

FREAKY FACT

I CAN CHANGE MY BODY COLOUR!

Did you know?

Roads that cross the crabs' migration route are closed while the crustaceans are on the move.

ROAD CLOSED
RED CRAB MIGRATION
NO ENTRY BY VEHICLES
BEYOND THIS POINT

DANGEROUS WATERS

These creatures could be from your nightmares.

BEWARE

Southern blue-ringed octopus

This octopus might be tiny, but it's one of the most dangerous creatures on Australia's southern coastline. It has one of the deadliest venoms known and its bite has killed people. Unfortunately there is no **antivenom** available. But the good thing is that it's very shy and likes to stay hidden away in rock pools. This little creature's common name comes from the small vibrant rings of blue that flash across its skin when it's feeling threatened or distressed. These warn predators that they are venomous and should be avoided.

STINGERS

FREAKY FACT

MY VENOM CAN KILL YOU!

Box jellyfish

This jellyfish produces one of the deadliest venoms in the animal world. It can kill a person in less than five minutes. Box jellies, as they're often called, appear along Australia's northern coastline from October to May each year.
And the best way to avoid them at that time is to swim only in special ocean enclosures that keep them out. Or wear protective clothing while in the water.

WEIRD INSECTS

Some Aussie flies create fairy wonderlands. Others eat poo. Many just drive us crazy!

FREAKY FACT

I GLOW IN THE DARK!

Glow-worms ▶

Glow 'maggots' would be a better name for these cave dwellers. That's because, even though they create beautiful fairy-light displays, Australian glow-worms are really fly larvae. And the common name for fly larvae is, of course, maggots. What's even creepier is these larvae are **carnivorous**.

At night a blue–green light produced by a chemical reaction glows on the end of these larvae. It attract insects to their sticky traplines. The larvae pull these up with their mouths to eat the captured prey.

◀ Australian bush flies

If you've ever been bothered by small flies that won't leave you alone when you're in the Aussie bush, chances are they're Australian bush flies. And they're probably females after your body fluids – like sweat, tears or even blood. The protein it contains helps them produce eggs. But when it comes time to lay their eggs, bush flies will be looking for something even more disgusting! The young of these flies – the maggots – develop in dung.

EXTREME MOTHS

Australia has up to 30,000 different types of moths. These are two of our most bizarre, for very different reasons.

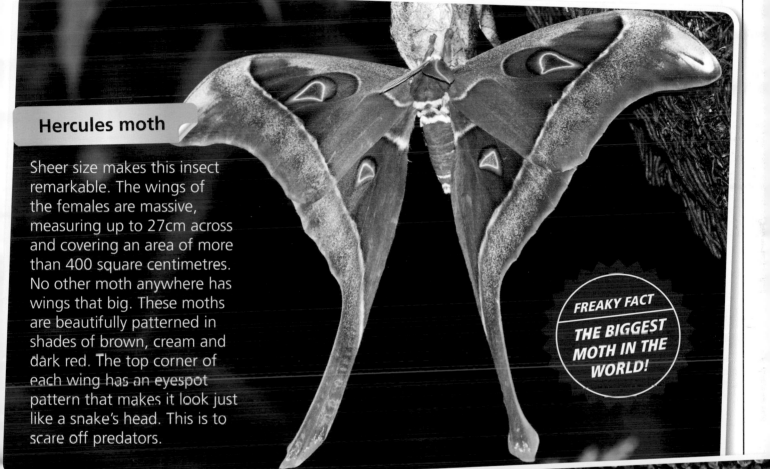

Hercules moth

Sheer size makes this insect remarkable. The wings of the females are massive, measuring up to 27cm across and covering an area of more than 400 square centimetres. No other moth anywhere has wings that big. These moths are beautifully patterned in shades of brown, cream and dark red. The top corner of each wing has an eyespot pattern that makes it look just like a snake's head. This is to scare off predators.

FREAKY FACT

THE BIGGEST MOTH IN THE WORLD!

Bogong moth ▶

Each spring, parts of south-eastern Australia are invaded by small hairy bogongs. For thousands of years, the adults of these night-flying moths have been making exactly the same **migration**. As the weather warms up they leave their winter breeding grounds in Queensland and New South Wales and head for the cool temperatures of the Australian Alps.

THE NOISIEST INSECTS

Our cicadas will set your ears ringing . . . and they have some of the most awesome common names of any insects.

FACT BOX

Song lures

The chirping calls produced by some species of Australian cicada can be so loud they can make your ears hurt! At least two large Aussie species – known as greengrocers and double drummers – can produce a noise louder than 120 decibels. That's about as loud as a jet engine and it's very close to the pain threshold for human ears. It's only male cicadas that sing and they do this mostly to attract females. Each different species has its own particular song.

Greengrocers

Greengrocer cicadas have distinctive red eyes and bodies that are most often coloured bright green. But this very common eastern Australian species also comes in other shades. Yellow ones are known as yellow Mondays. There are other, more rare forms: dark brown ones are called chocolate soldiers, turquoise-blue ones are called blue moons.

Q&A

Q: How long do Australian greengrocer cicadas live underground before coming up out of the soil and changing into adults?
A: About seven years.

Double drummer

This is Australia's loudest cicada. It's also our largest, with adults growing up to 15cm long.

Floury baker

While this cicada species is mostly brown, it gets its name because it looks like it's been dusted with white flour.

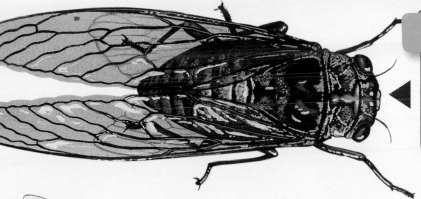

Did you know?

Although they don't live long as adults, cicadas often have very long lives. This is because young cicadas, which are known as nymphs, live underground and can remain there for years. Eventually they emerge from the soil and climb up trees to transform into flying adults. They leave behind the outer casings of their hard external skeletons on tree trunks.

Australian cicada-killer wasp

Australian cicadas have their very own specialist insect predator – the Australian cicada-killer wasp – and it's found wherever there are cicadas. It's a large wasp, about 4cm long, and the females have a creepy way of using cicadas to feed their young. When a female finds a cicada she paralyses it with her sting, then carries it back to an underground nest. There she lays an egg directly into the cicada's body. When her larva hatches, it feeds on the paralysed insect.

STICKY INSECTS

These weird-looking insects are some of our most unusual invertebrates!

Australian honeypot ant

Like most other ants and many bees, this ant species lives in colonies where groups of individuals have different responsibilities. The role of members known as 'honeypot' ants is to be living stores of food. Ants with the job of being 'honeypots' in a colony are known technically as 'repletes'. They are force fed by the worker ants to the point where the repletes' abdomens become fat with a store of sweet food. When this is needed by the colony, the workers stroke the **antennae** of the repletes and they spit up the nutritious sticky substance stashed inside them.

FREAKY FACT

I STORE HONEY IN MY BELLY!

FACT BOX

Lost and found

It was thought this stick insect had become extinct way back in the 1930s. But in 2001, 24 of them were found living on a rocky outcrop called Balls Pyramid, not far from Lord Howe Island. Scientists and wildlife officers mounted a massive rescue effort and some insects were recovered and sent to Melbourne Zoo to be bred. Thousands of these special stick insects have since been raised.

Lord Howe Island stick insect

It's sometimes called the tree lobster. But what this enormous stick insect really looks like is a black sausage with legs. Not so long ago, it was the rarest insect in the world. Fortunately, after one of the world's most successful animal conservation rescue efforts, its future looks safe.

FREAKY FACT
I CAN GROW UP TO THREE METRES LONG!

Giant Gippsland earthworm

We certainly grow earthworms big in Australia! This species has been known to reach a length of 3m, although most adult specimens are about one metre long. These worms are only found in one small patch of tall eucalypt forest in an area of just 100,000 ha, in a valley in south eastern Victoria.

FREAKY WORMS

WACKY WORMS

These odd forest creatures will make you squirm.

Velvet worm

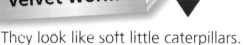

They look like soft little caterpillars. But velvet worms are **voracious nocturnal** predators that hunt other tiny creatures, like insects. They catch prey by tangling it in a sticky white fluid they spit from tubes on the sides of their head. This spit can reach as far as 30cm away: not bad for creatures that are about 2cm long. Tasmania has 20 of the world's 110 species of this weird 'worm'.

Giant blue earthworm

Very little is known about Australia's giant blue earthworm. This wriggling animal, which grows up to 2m long, is known only from the rainforests of remote Far North Queensland and is usually hidden away deep underground. When an animal is so brightly coloured it's often a warning to would-be predators that it's poisonous. But that's not the case here. To make things even weirder, this worm leaves a glowing trail of **bioluminescent** slime behind it.

Glossary

ambush	To hide and wait quietly then make a surprise attack.
antennae	Long, sensitive feelers on the head of an insect.
antivenom	Medicine for people who have been bitten by venomous animal that is made by milking venom from that species and diluting it.
bioluminescent	The emission of light by a living organism
carnivorous	An animal that eats meat.
forage	Look for food.
hibernate	Spend the winter in a safe place being dormant.
incubate	Keep eggs at an ideal temperature until they hatch.
migration	Travel from one place to another.
nocturnal	Active at night.
raptor	A bird that hunts its prey, often while in flight.
voracious	Having a big appetite and eating a large quantity of food.

FURTHER READING

Sand Swimmers: The Secret Life of Australia's Dead Heart
Narelle Oliver, 2013, Walker Books

Everything You Need to Know: Animals
Nicola Davies, 2013, Kingfisher

Animal Record Breakers
Steve Parker, 2013, Scholastic

National Geographic Readers: Predators Collection
2013, National Geographic